This book is dedicated to all who find Nature not an adversary to conquer and destroy, but a storehouse of infinite knowledge and experience linking man to all things past and present. They know conserving the natural environment is essential to our future well-being.

LAKE MEAD & HOOVER DAM

THE STORY BEHIND THE SCENERY®

by James C. Maxon

James C. Maxon retired from government service in 1990 after a career with the National Park Service and subsequently the Bureau of Reclamation. A graduate of the University of Denver, with a master's degree in archaeology from the University of Wisconsin, Jim lived and worked in the Lake Mead-Hoover Dam area for 20 years, gaining a thorough knowledge of and appreciation for the elements that constitute the vast Lake Mead country.

Lake Mead National Recreation Area, *located in southern Nevada and northern Arizona, was established in 1964. It includes Lake Mead (Hoover Dam) and Lake Mohave (Davis Dam).*

Front cover: Hoover Dam and Lake Mead, photo by Jeff Gnass. Inside front cover: Spring desert flowers, photo by David Muench. Page 1: Desert bighorn, photo by Gail Bandini. Pages 2/3: Temple Bar, Lake Mead, photo by David Muench.

Edited by Cheri C. Madison. Book design by K.C. Den Dooven.

Fourteenth Printing, 2005

LC 79-87573. ISBN 0-916122-61-1.

For eons a great river carved its way through the wilderness, a murky ribbon of churning, silt-laden water. Then came man with his gigantic dams and azure, sprawling lakes. Water, desert, dams, and mountains—astonishing contrasts in an immense land where nature, in the final analysis, is still in control.

The Lake Mead and Hoover Dam Story

The desert sun *embraces spiny cholla, blue lake, and endless desert ranges that stretch to the far horizon.*

DAVID MUENCH

Lake Mead National Recreation Area is more than just a story about water and water recreation. The area is a complex balance between the forces of nature and the forces of man.

The basic story is about too much water, at times, leading to flooding, even the loss of lives. The "something-had-to-be-done" answer was a monumental dam built in the early1930s.

The surrounding land is classic desert—hot, dry, seemingly barren. The two lakes in this National Park unit, Mead and Mohave, serve us all in many varied ways. The terrain that surrounds the lakes provides us with camping, hiking and the enjoyment of desert life. Make no mistake, there is a lot of life out in the desert. You just have to look for it!

The lakes provide recreation in the broadest of terms. Fishing, scuba diving, sailing, motor boating, house boat adventures—the list is almost endless. Even with over eight million visitors a year, there are coves to be found where you / your family can enjoy quiet and solitude. Again, you just have to look for it.

Lake Mead National Recreation Area was the first of its kind in our National Park System. It also represents the close cooperation between two agencies; The National Park Service and the Bureau of Reclamation. Each has their own mandates; yet working closely together they provide a seamless package for the betterment of all.

Water! Along with air, water is probably the most common element to sustain life. Water provides so much to our entire planet. Here at Lake Mead / Lake Mohave we see so much of what it does to make our life — Life!

Few activities bring one closer to the *transparent side of nature than water-skiing. Many people enjoy the gravity-defying exhilaration of gliding over the lake's surface.*

GAIL BANDINI

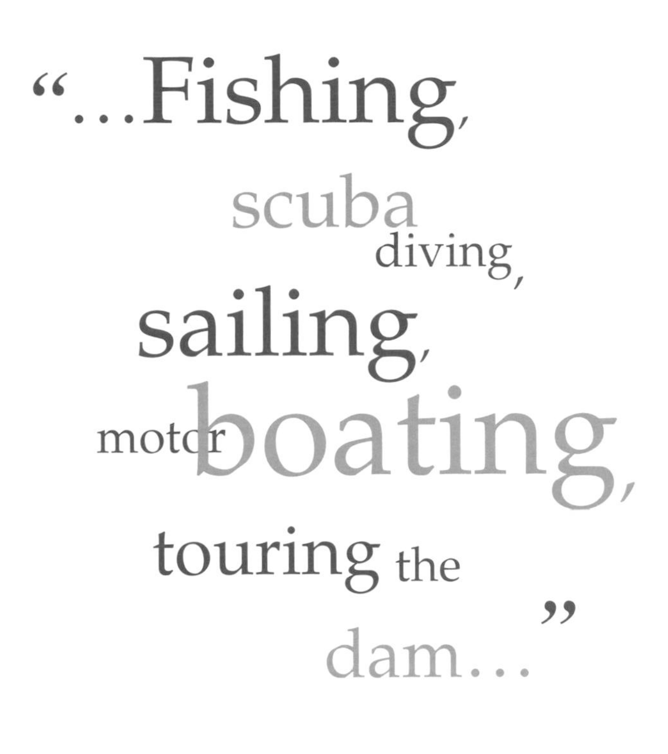

Hoover Dam, the world's largest hydroelectric installation *at the time of its construction, presented massive challenges to its designers and builders. Yet the project was completed in less than five years!*

GAIL BANDINI

"The lakes and desert vary in mood and color from moment to moment."

The River and the Lakes

Lake Mead, Lake Mohave: The two lakes man has created in the desert not only mark the taming of the Colorado River, they are themselves a recreational resource of national significance. Only recently a part of the Arizona-Nevada scene, the lakes nevertheless seem to be timeless elements in this desert setting. Their ages may be measured in only decades, but they represent the culmination of an intimate association between man and the river that has lasted for centuries.

Lake Mead was created first, with the building of Hoover Dam in 1935; Lake Mohave followed, with the completion of Davis Dam in 1953. Together they provide a setting for virtually all forms of water-oriented recreation. Warmed as it is by the desert sun, the water is a year-round resource enjoyed by millions of visitors from all over the world. Fishing, water skiing, sailing, scuba diving, swimming—all are frequent activities on the lakes. What could equal the thrill of landing a fighting bass, skimming the surface of the water on a pair of gravity-defying skis, or piloting your own graceful sailboat?

Beyond the 20th-century dams and lakes, the roar of the powerboats, and the splashing of swimmers near a rocky beach lies the great mystery of the ancient desert. There is time for serenity and reflection here. Time to enjoy the ever-changing beauty of the lakes and their mountain backdrop. Time to watch as the constantly changing patterns of light and shadow flow over the landscape. Time to wait out a sudden storm, and time to think about the enormity of the forces and the eons of time that shaped the desert.

DAVID MUENCH

A river cutting and twisting *its way through the desert landscape becomes part of a lake as man constructed Hoover Dam for flood control back in the 1930's. Beneath the calm surface of this "lake," the river continues to flow down to the sea, all the way to the Pacific Ocean. The water of Lake Mead and Lake Mohave provide a variety of benefits for people: enjoying recreational activities, electricity, water for farming, and year to year stability of water flow. All this from the Colorado River!*

The lakes and desert vary in mood and color from moment to moment. Each season brings its own changes—sometimes subtle, sometimes dramatic. If the gentle rains of late winter and early spring fall at the right time, the desert shows its gratitude by sprouting a springtime mantle of green splashed with brilliant flowers and grasses. The harsh summer follows, drying and desiccating, occasionally spewing forth torrents of water that rake the land. Gradually the searing summer heat gives way to the balmy, still days of autumn and winter, when once again the desert softens.

And always it rewards those patient enough to learn its secrets. For here, in this huge land of crimson desert and rugged mountains, sparkling water and towering sky, lies a depth of history, a history that is revealed in the shaping of the land by both man and nature.

JOSEF MUENCH

For eons the powerful Colorado, *unchecked and unhindered, cut its way through layer after layer of the Grand Canyon formations. Now, restrained in this area by the man-made dam and lake, the river is calm, especially here at Lower Granite Gorge, the lowermost section of the canyon. Shivwits Plateau is the highland area on the distant horizon.*

"Any story here must begin with the *river*..."

The Colorado River dominates the Lake Mead country. This says a great deal for the size and might of the river, for this is *big country*.

The river heads in the Rocky Mountains of Colorado, far to the north and east of the Lake Mead country. (It is joined in Utah by the Green River, which heads in Wyoming.) For countless centuries the river has scoured the land and cut great canyons with its thick, red waters (hence the *Rio Colorado*, a name bestowed upon it by the Spaniards, the first Europeans to see this country). The river was unpredictable. At times tens of thousands of cubic feet of the turbulent, muddy water roared through its canyons; at other times only a stream moved over its sandbars.

Before the dams were built, the river flowed westward out of the Grand Canyon into the desert before bending south. Now, lying at right angles to each other, the two lakes that were created by the dams stretch along the channel of the river for nearly 180 miles. Lake Mead covers the westward course of the river from the Grand Canyon to the bend where Hoover Dam is located; Lake Mohave stretches south below Hoover Dam for another 67 miles to Davis Dam. The Virgin River, which flows

GAIL BANDINI

Water, wind, and time have sculpted the vivid, *unusual sandstone formations of Redstone on the Northshore Scenic Drive of Lake Mead.*

***The black lavas of** Fortification Hill add a somber note to the brilliant, almost artificial colors of the Paint Pots on the Lake Mead shore.*

south into the Colorado at a point west of the Grand Canyon, has also been affected by the creation of the dams and the lakes. The Overton area of Lake Mead now backs up about 25 miles of the Virgin to the north.

The Lake Mead country is as varied as it is big. Elevations range from nearly 7,000 feet above sea level on the Shivwits Plateau north of the Grand Canyon to just over 500 feet above sea level at Davis Dam. Elevation changes are accompanied by changes in climatic conditions and types of wildlife. The variations are unusual: the ponderosa pine, the roadrunner, the mule deer, and many types of cacti are all part of the Lake Mead scene. Through it all, the river—and now the lakes—provides a bond that brings a measure of continuity to this vast region.

The river shaped much of the landscape and influenced much of the history of the Lake Mead country. Yet it is "only" a few million years old—an upstart, geologically speaking. Events that date from the very beginning of the earth set the stage for the birth of the river. The story of the river, then, starts with a look into those primordial times. It can be only a brief glimpse; so little is known of that incredibly ancient era here that we can do little more than note the presence of the rocks—and wonder.

Rocks from the earliest era of the earth's history outcrop at several places within this region. The schists of Saddle Island and the metamorphic rocks of the Newberry Mountains west of Davis Dam are silent reminders of that era, a time that predated all life.

The appearance of life is recorded here too, marking the end of an era that lasted for eons. Fossils of the earliest life forms are contained in

Boaters explore *Surprise Canyon, a side-canyon hideaway. High lake levels enable boaters to travel into parts of the lower Grand Canyon that are otherwise nearly inaccessible.*

rocks of the *Paleozoic* era (600 million years ago) exposed in the Muddy Mountains, the rugged desert hills that lie northwest of the angle formed by the Overton Arm and the Boulder Basin of Lake Mead. (The hills themselves were the result of more-recent mountain building.) The Paleozoic rocks are mainly sediments deposited in an ancient sea at a time when the Lake Mead country was part of a long north-south trough. This trough, called the *Cordilleran geosyncline*, stretched through much of what is now the western Great Basin region. The Paleozoic era ended in this area as the land rose and the seas receded about 200 million years ago.

Desert replaced sea here during much of the succeeding geologic era (the *Mesozoic*). At times, winds howled relentlessly across the ancient deserts and formed great sand dunes. Fossil remnants of these dunes make up the colorful red-sandstone formations exposed in Nevada along the Northshore Scenic Drive of Lake Mead and in the Valley of Fire Nevada State Park.

Near the end of the Mesozoic, a hundred million years ago, earth movement—generated by forces of almost incalculable strength—began. For eons they twisted, warped, smashed, and sheared the rocks. In one area, for example, an entire block of the older Paleozoic rock was gradually pushed up, grinding its way over the top of the rock remnants of Mesozoic sand dunes. The result was that billions of tons of rock were moved at least 13 or 14 miles; this was the giant *thrust fault* that formed the Muddy Mountains.

Then came the volcanoes. Masses of molten rock pushed up through the older sediments. Here and there it broke through the crust of the earth and tons of the fiery *magma* were disgorged, becoming *lava* at the surface. Fortification Hill—the great, black, flat-topped mountain that guards the lower basin of Lake Mead—is capped by layer upon layer of the lava (now *basalt*) from some of those volcanoes.

About 70 million years ago, continued volcanism and earth movement ushered in the most recent era, the *Cenozoic*. At least three distinct series of periods of volcanic rocks of the Cenozoic are traceable here. The centers of that volcanic activity appear to be located in the areas flanking Lake Mohave south of Davis Dam.

Faulting, or shearing, and moving of blocks of the earth's crust also continued. The north-south series of mountains and valleys of the Lake Mead country were formed during this time. Along north-south fractures, great blocks of the earth were alternately forced either up or down in relation to each other. Raised blocks became mountains; lowered blocks became valleys. One of these troughs, now the site of Lake Mohave, was there for the river to follow, once it appeared.

The River Appears

So the stage was set for the river. But where did it come from? Geologists have more than one theory about the origin of the Colorado, but the one most generally accepted is that known as the "ancestral Little Colorado theory." Greatly simplified, it goes like this:

Some 10 million years ago, the ancestor of the Colorado, the "Little Colorado," flowed from the Rockies into central Arizona, emptying into a lake (long since vanished), which was bounded by a highland to the west (the Kaibab Plateau). The stream that drained the west side of this highland gradually cut headward to the east. Eventually the flow of the ancestral river was diverted into this drainage to the west, and the Colorado River was born. The highland continued to rise and the river continued to slice through it, maintaining its new westward course and eventually carving out the astonishing gorge that we know today as the Grand Canyon. (The story of the remarkable events behind the creation of the canyon is told in a companion book, *Grand Canyon: The Story Behind the Scenery*.)

There is some evidence that nature wrestled with the river and held it in control for a time long ago—about 2 million years before man harnessed it. The extensive lake-deposited sediments that dot the present valley of the Colorado River are thought to be remnants of a huge natural lake perhaps a thousand feet deep in places. The lake, known as "Lake Chemehuevi," was perhaps the result of a natural barrier that occurred when lava flows or natural landslides dammed a deep, narrow canyon, location unknown. This great dam held the Colorado in check for hundreds, perhaps thousands, of years—long enough for the Chemehuevi deposits to be formed before the river breached the natural barrier and continued its relentless course to the sea.

During the millions of years in which the Colorado was scouring out the Grand Canyon, it also carved Boulder Canyon, Black Canyon, and dozens of other canyons in the Lake Mead country and beyond. Even today the forces that shaped the land continue to work. The river, where it is not dammed, and its tributaries wear away their canyon walls; wind and rain wear away the high surfaces and fill in the low areas. But they do it usually with such slowness that the changes are almost imperceptible to man limited as he is by his short time frame.

Yet sometimes the forces that shape the landscape are sudden and violent. Many earthquakes (minor ones) occur in the area, indicating that the crust of the earth will continue to shift and warp. In the fall of 1974, a flash flood, caused by a cloudburst in the Eldorado Mountains, poured thousands of cubic yards of mud, rock, debris, and angry water into Lake Mohave, taking lives and totally destroying the recreational development known as "Nelson's Landing." Such are the reminders to man that he must heed these powerful forces and take measures to guard against them.

The last several million years of the geologic story here were marked by the ice ages, a time when great continental glaciers formed and the climate of the earth cooled. The glaciers covered thousands of square miles—about a third of the earth surface—but they didn't reach this far south. Associated climatic changes, however, did affect this area. During the glacial advances, the climate here was cooler and more moist than it is now. Many of the low, dry basins that we see today once held permanent lakes fed by free-flowing streams.

During the last of these cool, wet periods—perhaps no more than 10 or 15 thousand years ago—a new element appeared on the scene, one of nature's agents that was destined to change the landscape drastically and perhaps irreversibly. Its name was *man*.

CLIFF SEGERBLOM

Desert, lake, and rock are immersed *in the rich tides of warmth released by a desert sunset. A beautiful sunset is an almost everyday occurrence that most desert dwellers take for granted.*

SUGGESTED READING

DOTT, ROBERT H., JR., and BATTEN, ROGER L. *Evolution of the Earth*. New York: McGraw Hill Book Co., 1971.

LONGWELL, CHESTER R.; E. H. POMPEYON; BEN BOWYER; and R. J. ROBERTS. *Geology and Minerals Deposits of Clark County, Nevada.* Nevada Bureau of Mines Bulletin 62. Reno: University of Nevada, 1965.

PRICE, L. GREER. *Grand Canyon: The Story Behind the Scenery.* Las Vegas, Nevada: KC Publications, Inc., 1991.

"...if the early hunters of the Southwest did roam the Lake Mead country, they surely would have known the river, which had been relentlessly carving its channel for eons."

Those Who Came Before...

It is no accident that virtually all human history in this great, dry land focuses upon the Colorado—a thread of muddy, turbulent water that rushes through a labyrinth of canyons and mountains. The river was a magnet that drew all of the various peoples who lived here at one time or another. It became the source of life for the dusky-skinned Indians who planted fields of corn along its banks and for the stoic Mormon farmers who came along afterward and tilled the same land. And it was the lure of the river and the beaver along its banks that brought the mountain men, the trappers. The river was the travel route for early explorers and steamboat captains alike, indeed for nearly all who chose to come to this area.

The Colorado River has always been a crossroads for humans, whether they stayed for 10 thousand years, as the Indians did, or simply passed through, as did the "forty-niners" on the overland trek to the California gold fields. Thus when we begin to know and understand the river we can begin to know and understand the players in the human pageant that for centuries has been played out along the banks of the river and in the desert beyond.

For the desert, too, is part of the stage in this remarkable drama. Here, perhaps more than in any other place, humans have been affected by their environment. The harshness of these surroundings forces those who inhabit it to come to grips with the reality here or be utterly defeated by it. Even so, the vast desert is not as barren as it seems. It holds a variety of minerals, plant and animal life, and there is always, of course, the breathtaking beauty of its ever-changing landscapes. All of these factors have profoundly influenced the attitudes and shaped the actions of the people who have lived here.

DAVE HUNTZINGER

Petroglyphs—As in the *past, pecked and etched images on boulders and cliff faces have great significance for Native Americans today, and intrigue thousands of visitors who see them.*

DAVID MUENCH

Water, water, water! It carves mountains. *It creates the river. It gives life to the desert cactus and flowering brittle bush. Why here and not ten feet to the right or left? Where water collects—even in minute quantities, desert plants flourish.*

The First People

Before the end of the Ice Age, small bands of hunters appeared in the Southwest. These groups roamed the area in search of mammoths, horses, and camels, now extinct, and animals more familiar to us, such as deer and bighorn sheep. Campsites in the Southwest, or more often the sites of a hunt, have been carefully excavated and researched, and the evidence assessed. At such a site, one can sense the drama in an imagined scene in which a few prehistoric hunters armed with stone-tipped spears bring down some huge, woolly mammoth in a thunderous collapse.

It seems fairly certain, then, that humans were in this general region at the end of the Ice Age. But did they actually hunt *here*, in the Lake Mead country? Tantalizing clues found at Tule Springs north of Las Vegas, Gypsum Cave just beyond the north shore of Lake Mead, and Rampart (Sloth) Cave on the Colorado just outside the recreation area, point to just such a possibility. These areas have collectively yielded remains of late Ice Age animals—mammoths, horses, and camels—as well as mountain goats, marmots, and a desert tortoise that must have been carried to its location either by humans or a large type of prehistoric hawk. Gypsum and Rampart caves have also produced exciting remains of the giant ground sloth.

Archaeologists who worked in the Tule Springs and Gypsum Cave areas during the late 1920s and 1930s believed they had found definite evidence that humans were here 10 to 15 thousand years ago. There is some doubt as to these early interpretations, however, so it seems that for the present, at least, we can only speculate. But, if the early hunters of the Southwest *did* roam the Lake Mead country, they surely would have known the river, which had been relentlessly carving its channel for eons.

The retreat of the glaciers to the north and the emerging desert conditions marked the end in the Lake Mead country of the presence of the great animals of the Ice Age and other life that had adapted to the cold environment. The evidence that humans existed in this area from about this time on is firm.

As desert conditions began to predominate, people's life style changed to accommodate them. The transition was gradual, spanning generations. Small desert animals, even rodents and reptiles, became increasingly important as food sources. Plants became dietary staples, too, as the people learned to harvest the fruits of such unusual plants as yucca, cactus, mesquite, and agave. This new way of living, known to anthropologists as the *Desert* culture, developed around the utilization of a variety of desert plants and animals as they became seasonally available.

The Paiutes

So perfectly did the Desert culture adapt to the environment that, once established, it remained virtually unchanged. The Paiutes whom the first white explorers encountered along the river a hundred and fifty years ago represented the culmination of the 10,000-year-old desert life style. Except for minor variations, the lives of the Paiutes have changed but little from their prehistoric progenitors.

The Paiutes traveled in small, family-related bands, but this was far from aimless wandering. Their routes were well defined. They knew the location of each grove of mesquite, concentration of agave, and source of water away from the river. With an almost unerring sense of direction, they searched out plants and animals at the proper places and times for harvesting and hunting.

RUSS GRATER

Giant sloths roamed the Lake Mead area late in the Ice Age. Rampart (Sloth) Cave, in Lower Granite Gorge, Grand Canyon (just outside the recreation area), has produced the finest remains of this extraordinary animal that we have today. Paleontological finds there—including a two-foot section of skin with the rich, reddish-brown hair still intact—have enabled us to re-create a reasonably accurate image of the sloth, such as this sketch by artist Albert Long.

The Indians utilized every resource the desert *offered. The yucca provided materials for mats, sandals, nets, baskets, and rope. Its cucumberlike fruit was an important food source in the spring.*

DAVID MUENCH

In the spring, yucca, agave, and cactus plants growing in the higher desert areas yielded fruits, which were eaten fresh or were dried, roasted, and ground into meal. During the summer, tuberous desert plants such as the sego lily became a source for food, and in late summer the Indians harvested ripe mesquite and catclaw beans. During the fall, small bands of Paiutes gathered in the high desert mountains, the women to harvest pinyon nuts and the men to hunt mule deer and bighorn. This was a time for socials and rituals, perhaps the only occasions during the entire year that the peoples of different bands could meet and have a chance to renew old acquaintances and visit with their kin.

The key to the survival of these Indians under what we might look upon today as impossible conditions lay in their ability to use a variety of resources. They knew that no single desert resource could successfully sustain human life. Many food sources were utilized, even those that were available in only small quantities. The fact that the Paiutes were eminently successful in creating a way of life out of very little is undeniable: It was not until the 1870s, when these Indians were placed on reservations, that their ancient Desert culture way of life began to disappear.

The Mojave

Warriors...farmers...traders...dreamers, the Mojave represent a rich and complex people. Traveling widely throughout the Mohave Desert and as far north as the Grand Canyon, the Mojave raided neighboring peoples and participated in complex and shifting political alliances. Through extensive trade networks, they exchanged corn and other crops for upland game and shells from the

RUSS GRATER

In this painting of the "Lost City" by artist L. A. Ramsey, Pueblo farmers greet the morning sun. *During the seven centuries preceding A.D. 1150, when these agriculturalists mysteriously abandoned their villages, they farmed the fertile floodplains of the Virgin and Muddy rivers. Their ancient villages now sleep far beneath the waters of the Overton arm of Lake Mead.*

Pacific Ocean. This finely-tuned society was adapted to both desert and river life, with the Colorado River and all its landmarks as their center.

The Mojave and other Yuman-speaking tribes farmed the banks of the river. Annually the river, swollen with spring melt water from the Rockies, spilled over its banks and deposited rich silt on the bordering lowlands. Corn, pumpkins, squash, and cotton contributed as much as 50 percent to their diet. The Mojave also collected staples like honey mesquite pods and other wild plants from the washes and hillsides. Squawfish and other fish from the river supplied needed protein, as did wild game obtained by small traps, snares, and bows and arrows.

The Mojave constructed simple tools and housing, but ones that were well-suited for the area. Pottery, knives, and scrapers tended to be functional and utilitarian in form. Summer housing consisted of small brush ramadas and other informal structures clustered in small hamlets, or rancherias, along the river. More substantial earthen and wood homes sheltered the people in cooler months.

Contrasted with their simple material world, the Mojave social and ceremonial world was intricate, intertwining into all facets of daily life. Elaborate song cycles, sometimes taking days to recite, told of their origins and documented ceremonial migrations to be traveled in both dreams and on foot. Rituals, often involving elaborate tattooing or the pecking of petroglyph images, punctuated calendar events, rites of passage, and other important events. Further south, the Mojave and others scratched and tamped large earth drawings or *intaglios* into ancient desert pavement. They revered mountains and other landmarks along the river as sacred. Connecting both cosmology and subsistence were ceremonial paths marked by stone shrines and carefully recorded in dream-songs. Such paths and dreams defined a way of life for these warrior-farmers.

This life style was interrupted only when the explorers and miners who came to the river in search of wealth and adventure happened upon the Mojave villages just over a hundred years ago.

The Pueblos

By the beginning of the Christian era, several agricultural societies much more sophisticated than the Paiutes and Mojave were emerging in the Southwest. One of these—the *Anasazi* (or Pueblo) culture, centered in the Four Corners area (where Arizona, New Mexico, Colorado, and Utah converge)—successfully pursued both dry farming and irrigation.

This way of life was so successful that it spread west across Arizona into southern Utah and the Lake Mead area. The Colorado provided the water to nourish the crops of the Pueblos; thus the river and its tributaries became the very source of their existence. By A.D. 500, an efficient farming society had developed along the Virgin and Muddy rivers, just above their junction at the Colorado.

NORTH WIND PICTURE ARCHIVES

"Mountain men" seeking valuable beaver pelts were *drawn to the banks of the lower Colorado River during the 1820s and 1830s. Period illustration by Frederic Remington.*

For seven centuries these people succeeded in irrigating their crops of corn, beans, and squash by diverting river water.

Far removed from the cultural center to the east, these farmers might well be called the "country cousins" of the Anasazi. This is not to imply that they were in any way inferior. Evidences of their arts and crafts and adobe-walled apartment houses reveal a way of life nearly as complex as the Anasazi and—for that matter—more sophisticated than many a modern civilization.

Most of the significant archaeological sites of the Pueblo culture—the "Lost City" pueblos and the Virgin River salt quarries—were inundated by the Overton arm of Lake Mead. Studies of these villages made prior to that time, however, suggest that these people left the area about A.D. 1150. The reasons for the abandonment of the Pueblo villages are shrouded in mystery. Perhaps a drought, pressure from other, more aggressive peoples, or a combination of factors drove them out. The only thing certain is that the life they had so long maintained with the river and desert was somehow suddenly interrupted. Whatever the catastrophe or change that befell them, these people obviously felt they could no longer hold on to their agricultural way of life in this particular area.

At the Lost City Museum in Overton, Nevada, is a reconstruction of an ancient village that was covered with water when Lake Mead formed. The museum also displays hundreds of artifacts that bring the Pueblo people vividly to life.

The First Explorers

Autumn had warmed the barren hills, and only the dense vegetation along the placid river seemed to escape the parching effects of the sun.

It was October 5, 1826. A small party of buckskin-clad mountain men traveled along the banks of the Virgin to the north, toward the Colorado. Jedediah Smith and 15 others were ostensibly looking for new areas to trap beaver. Yet who would doubt that the lure of this unknown country was as least as strong as the lure of furs. Smith's journal entry for this day is a masterpiece of understatement:

> *"I followed Adams [Virgin] river two days further to where it empties into the Seedskeeder [Colorado] . . . and went down it four days a south east course; I here found the country remarkably barren, rocky, and mountainous; there are a good many rapids in the river, but at this place a valley opens out about 5 to 15 miles in width, which on the river banks is timbered and fertile. I here found a nation of Indians who call themselves* Ammuchabas *[the Mojave]; they cultivate the soil, and raise corn, beans, pumpkins, watermelons, and muskmelons in abundance, and also a little wheat and cotton. I was now nearly destitute of horses, and had learned what it was to do without food. . . ."*

Historians have been unable to trace this route through the Lake Mead country, the first time it had been traversed by white men. We do know, however, that Smith traveled from the Great Salt Lake down the Virgin to its confluence with the Colorado and then followed the Colorado southward to the Mojave villages near the present site of Needles, California. Whether he was forced to detour around the thousand-foot chasms of the Boulder and Black canyons or was able to thread his way across them in their low autumn flows is impossible to determine.

Smith's stay with the Mojave was a friendly one (later encounters, by other trappers, would not be as amicable), after which the party journeyed on and eventually reached the Mission of San Gabriel at Los Angeles.

Some 50 years earlier, Father Garces, a Spanish missionary priest, journeyed up the river to the Mojave villages. About this same time, Father Escalante crossed it far to the east. Others followed, including Antonio Armijo, a Mexican trader from Santa Fe, who in 1830 would pioneer much of the route of the Spanish and Mormon trails to Los Angeles. But to Smith goes the distinction of having been the first to follow the river through the Lake Mead country and record his findings.

By the early 1850s the United States Government recognized the need to establish a series of military posts to provide assistance and protection for the growing number of immigrants journeying to California and for the settlements springing up along the route and the river itself. Enterprising men turned to the river as the quickest and most logical way of transporting people and supplies from the coast to the interior settlements.

Steamboats on the Colorado

The first steamboat on the Colorado was the *Uncle Sam*. A sidewheeler, brought in sections from San Francisco and assembled at the mouth of the river, it was 65 feet long and 16 feet wide. It reached Yuma on December 2, and operated on the lower river for almost two years. The boat sank in Pilot Knob on June 22, 1854.

The steamers and their captains provided some of the most picturesque scenes in the pageant that had been unfolding on the river since humans arrived. The boats themselves were patterned after the larger eastern river steamers of the period and, although designed to operate in the often shallow flows and ever-shifting sandbars, in practice they often ran aground. Both side- and stern-drive paddle-wheel types of boats were used. The largest of these, the sternwheeler *Gila*, was 149 feet long. Riding in a stifling passenger cabin of one of these steamers and watching as the captain coolly extracted his boat from a sandbar by cutting through the obstruction with the paddle wheel must have been a never-to-be-forgotten experience.

By the mid-1850s, several small paddle-wheel steamboats were plying the Colorado from the Gulf of California up to Yuma. Captain George Johnson and his Colorado River Steam Navigation Company had a virtual monopoly on commerce on the river. It was Johnson who convinced the government to finance an exploratory trip up the river to determine its upper limits of navigation. The assignment he coveted went, however, to a young West Pointer, Lieutenant Joseph C. Ives.

***Although steamboats on the Colorado seem an** incongruous idea to us now, they were actually a vital element that for almost half a century (from the 1850s until the early 1900s) linked river settlements and the outside world. Typical of the shallow draught steamers that braved the ever-present sandbars and wild currents of the Colorado River, the picturesque* Cocopah *and its "Barge No. 3" prepare to do battle with the rapids of Black Canyon. By attaching cables from steam-driven winches to iron ringbolts anchored to the canyon walls—as depicted in this painting by Francis H. Beaugureau—the steamers negotiated the white-water stretches. Despite irritating delays, they were usually successful in reaching settlements upstream.*

DAVE HUNTZINGER

The infuriated Johnson determined to beat Ives and accomplish his own investigation. Able to get Lieutenant James White and 16 soldiers from Yuma as escorts (because of the Mormon war scare and Indian unrest), Johnson and another 16 civilians set off from Fort Yuma on December 31, 1857. It was White who provided the description of the journey in his brief but accurate report of the river covering the 320 miles above the mouth of the Gila. By the time Ives had launched his little steamer, the *Explorer*, on January 11, 1858, Johnson was far ahead and proceeding northward in his side wheeler, the *General Jesup*.

Ives' *Explorer* had been built in Philadelphia and shipped in pieces by boat to the Colorado River delta, where it was reassembled. A clumsy, iron-hulled affair, the *Explorer* ran aground a mile above camp. It took five hours to extricate it, much to the delight of onlooking Indians. But despite progress that was sporadic and many time-consuming episodes with the exasperating sandbars so much a part of the river, the *Explorer* reached the

NPS PHOTO

mouth of Black Canyon in March, well above Johnson's reported mark. There Ives once again ran aground, this time with extensive damage. While repairs were being made, Ives and two others took a skiff up through Black Canyon, reaching the mouth of "Las Vegas Wash" (which they identified as the Virgin River) before returning.

Meanwhile Johnson had started his return (January 21) and had encountered Ives' party going upriver. The *General Jesup* was soon to run out of luck, however; 35 miles from the fort it hit a snag and sank ignominiously.

The honor of being first belongs to Johnson, who later made further but unsupported claims that he reached as far as Eldorado Canyon. It was Ives, however, who had official sanction, and due no doubt to the prompt and thorough report that he published in 1861, he is the one most often associated with the first exploration of the Colorado by steamboat. (White's report was also well documented, but it was not publicly printed until 1955, nearly a century later.)

An interesting sidelight of Johnson's return journey was the chance meeting with Lieutenant Edward F. Beale and his "Camel Corps." Beale was the instigator of the experimental use of camels in the Southwest in finding a "wagon road" from Fort Defiance, New Mexico, to the Colorado. Equipped with about 25 Arabian camels, a few native camel drivers, a military escort, and supplies including 350 sheep for food, Beale started his strange venture. He didn't believe warnings that camels could not swim. As a test, he herded into the water a strong male camel who, when "he found himself out of his depth struck off without hesitation for the opposite shore, swimming high and with perfect ease." It was after having wintered his camels in the Sierra and on the return trip to New Mexico that Beale made his astonishing encounter with Johnson, who helpfully ferried Beale's expedition across the river in the *General Jesup*.

"...with about 25 Arabian **camels**... Beale started his **strange** venture."

Even though the camel experiment was successful to some degree, other events—such as the onset of the Civil War—intervened, and the idea was never fully applied. The demise of this curious experiment ended a colorful chapter in the history of the Southwest. In subsequent years remnants of this unique corps were encountered throughout the Southwest, giving rise to exaggerated stories and speculations involving "wild camels" by those unaware of the little-known experiment.

GAIL BANDINI

***Boaters are often privileged to observe** desert bighorn sheep on cliffs overlooking the lake. The largest desert mammal, bighorn will not live more than two miles from permanent water. On steep rocky outcrops they can easily outrun their enemies.*

Ives and Johnson were followed by other steamboat men. There was a short period when these captains would negotiate the rapids in Black Canyon by literally hauling or lowering their boats through the rapids. Cables were attached to heavy iron ringbolts anchored to the canyon walls; steam-driven capstans or winches on the boats supplied the power to play out or haul in the cables.

The steamers lasted on the river until the beginning of the 20th century, but the coming of the railroads 20 years earlier had marked the beginning of the end of the paddle-wheel era on the Colorado.

Gold Fever

Ives, Johnson, and many other steamboat captains established the feasibility of operating steamboats above Yuma. But it was gold that provided the incentive. It had been discovered in the river at the mouth of Eldorado Canyon, and sources of the placer (free) gold were quickly traced. By the early 1860s a miniature "forty-niner gold rush" was developing at Eldorado. Settlements sprang up on the Colorado and in the Techatticup Wash to the west. Business boomed, and for a time Captain Johnson moved the headquarters of his steamboat company to Eldorado.

The gold fever born at Eldorado spread up and down the river. New discoveries followed. Placer gold was taken from Gregg's Basin on what is now the upper end of Lake Mead. For a time, a "stamp" mill was operated on Cottonwood Island (now flooded by Cottonwood Basin on Lake Mohave). Ore from the minesat Searchlight, Nevada, 14 miles to the west, wastransported to the mill via a narrow-gauge railroad. As late as the 1920s, mines in the Katherine area produced gold and other metals. A few mines in the Lake Mead country continued to produce even up until the beginning of World War II.

Government Explorers

Jedediah Smith and others had pioneered the routes along the river. But the frenzy generated by the discovery of gold in California meant that easier and quicker routes must be found. Quite naturally the government became interested and sent explorers to the Colorado country.

In the fall of 1851, one of these, Lieutenant Lorenzo Sitegreaves, and his small party set out from Santa Fe to explore a route to Los Angeles that might prove more direct than the circuitous Spanish Trail pioneered by Armijo 20 years earlier. Sitegreaves' journey across New Mexico and Arizona (which approximated the present route of U.S. Highway 66) was long and arduous; men and animals suffered terrible thirst and hunger. When Sitegreaves and his bedraggled band of soldiers and Mexican mule-skinners came upon the Mojave villages sprawled along the banks of the Colorado near the present-day Davis Dam, the sight must have been an extremely welcome one, to put it mildly. The stay with the Mojave was less than pleasant, however. In fact, there was bloodshed. Even so, Sitegreaves was able to trade for urgently needed supplies and the party eventually safely reached the missions at Los Angeles.

The Colorado attracted others by its size, grandeur, and awesome power. They came to conquer it, but soon learned that the struggle it would exact from them would be of heroic proportions. The river established its own terms, and those who came to learn its lessons firsthand would never forget them.

The Powell Expedition

Of the dozens who came to test the river, one name stands out more clearly and heroically than all the rest, for he was the *first* to test the will of the river. John Wesley Powell—scientist, geologist, teacher, government explorer, one-armed Civil War veteran—set out in May, 1869, to explore the still-uncharted upper reaches of the Colorado. From its beginnings in the Rockies, and throughout the Grand Canyon, its course was a mystery heightened by stories of tremendous waterfalls, underground chasms, and swirling whirlpools that, it was said, could destroy men and boats in moments.

Powell meant to learn the truth for himself. He and nine other men put into the river in four wooden boats at Green River, Wyoming. In the course of the journey, Powell and his men found heartbreaking misfortune and backbreaking toil. The danger of the treacherous rapids and the hardships of constant hunger, weariness, and worry were always with them. One boat capsized. The Englishman Goodman, his nerve shaken by the loss of the boat and its load of food and valuable equipment, left the expedition in July, before the men plunged into the point of no return.

On August 27, 1869—after weeks of fighting some of the Grand's most violent rapids—the scenario is this:

Three of the men (Captain Howland, Howland's brother, and Dunn), their confidence in Powell waning, decide that the only chance for survival lies in climbing out of the canyon and striking out for the Mormon settlements that they believe lie somewhere to the north. Powell, having talked with Captain Howland and beset with doubt and worry for the safety of all his men, reexamines their position and reaffirms his calculations. In his words:

> *As soon as I determine all this, I spread my plot on the sand, and wake Howland, who is sleeping down by the river, and show him where I suppose we are, and where several Mormon settlements are situated.*
>
> *We have another short talk about the morrow, and he lies down again; but for me there is no sleep. All night long, I pace up and down a little path, on a few yards of sand beach, along by the river. Is it wise to go on?*
>
> *. . . I almost conclude to leave the river. But for years I have been contemplating this trip. To leave the exploration unfinished, to say that there is a part of the canyon which I cannot explore, having already almost accomplished it, is more than I am willing to acknowledge, and I determine to go on.*
>
> *I wake my brother, and tell him of Howland's determination [to leave the expedition] and he promises to stay with me; then I call up Hawkins, the cook, and he makes a like promise; then Sumner, and Bradley, and Hall, and they all agree to go on.*

This old engraving captures the drama of that *first incredible journey, when Major Powell and his men, in 1869, challenged and survived the awesome Colorado. After three months of hair-raising adventure on the river, they landed peacefully near what is now Echo Bay on Lake Mead.*

POWELL'S 1875 REPORT

The following morning the two groups part, each thinking the other is doomed. Howland's group successfully climbs out of the canyon and reaches the Shivwits Plateau, only to be killed by Paiutes, probably mistaking the bearded, ragged, and gun-toting trio for "bad miners" turned out of Diamond Creek, a roaring mining camp across the river. (To the Indians, all white men looked the same.) On the river, one boat has splintered so badly it has to be abandoned. Powell and his men safely run Separation Rapids with the remaining two boats.

Two days later, Powell's epic exploration of the river plays out its final scenes. Under the hot, late-August sun, at the place where the Virgin River enters the Colorado, the six remaining men of the expedition are hailed in greeting by Mormon settlers. They have struggled with the river for three long months.

The Colorado was a formidable foe, but it had met a worthy adversary. Powell's strength and perseverance proved once and for all that the river would, albeit grudgingly, allow humans to travel its course.

"The Colorado was a formidable foe, but it had met a worthy adversary."

In 1871 Powell came back again to explore and make a full survey of the river and its surroundings. Others came too. George M. Wheeler, another government explorer and something of a rival of Powell's, led several successful expeditions into the western states. In 1871 he toiled up the river with rowboats to Diamond Creek. In 1890 Robert Brewster Stanton completed the first run of the Grand Canyon in almost 20 years, having been chosen to head a survey of the canyon to determine whether a railroad route was feasible. (Stanton reported that it was, an astonishingly optimistic conclusion, given the facts of his harrowing and disastrous journey.)

Today thousands travel the Colorado through the Grand Canyon in search of adventure and recreation. Their experiences have often been reported in grand and glowing terms. But no trip can ever equal Powell's. His was the first.

The Mormons

Not all of the latecomers to the river were seeking adventure, scientific information, or mineral riches. Some came to use its rich soil and water to feed and clothe their families, just as the Indians had done many years before.

By the 1860s, the Mormons, who had settled the Salt Lake area 20 years earlier, were looking for new areas to colonize and establish missions. Of particular importance was the need to find a reliable route to the West Coast for purposes of travel and commerce.

As others had found before them, the fulfillment of this need seemed to lie with the river. The success of Captain Johnson's steamers pointed to a more reliable alternative to the tortuous overland route of the Spanish. A steamboat port established on the Colorado could provide a logical link that would connect the overland route to the route between Salt Lake and the new "Dixie" settlement (named because of its cotton crops) on the Virgin River in southern Utah. A land-river route would eliminate the tortuous desert leg of the trail from Salt Lake to the West Coast.

Elder Anson Call and a small band of the faithful were sent south to provide such a steamboat port on the river and to establish farming colonies in the Moapa Valley. Arriving late in 1864, Call's people set out to construct their buildings—including a sturdy, stone warehouse—to serve the port of Callville. Black Canyon proved, however, to be a formidable obstacle, despite the optimism and relative success of the steamers on the lower river. The steamers experienced such difficulty in negotiating the rapids that a regularly scheduled service above Eldorado Canyon was out of the question, and completion of the Transcontinental Railroad in 1869 made further efforts unnecessary. Callville's brief existence came to an end. Today ski boats skim over the flooded ruins of Anson Call's stone warehouse.

Although the steamboat venture had failed, the colonization of the Moapa Valley continued. By the late 1860s the Mormons were farming the fertile bottomlands tilled 700 years earlier by Pueblo farmers. St. Thomas, West Point, and other bustling farming communities sprang up. But problems from an unexpected source posed new difficulties. The eastern boundary of the newly formed state of Nevada was shifted eastward, as a result of an 1870 survey, and the farmers found themselves facing state tax assessments that for the most part they could not pay. Sadly they packed their few possessions, abandoned their homes and fields, and headed back to the Dixie and Salt Lake areas.

One family stayed on—that of Daniel Bonelli, an obstinate Swiss. Bonelli possessed, in addition to tenacity, great ambition and a good head for business. He succeeded in making a living by pioneering a ferry service across the river where the Virgin flows into the Colorado and by mining the salt deposits that the Pueblo people had discovered. Place names such as "Bonelli's Landing," "The Temple," and "Rioville" come from this era.

By 1880 the Mormon people were once again settling in the Moapa Valley. The villages of Overton, Logandale, and Glendale date from this period. Today the neat, prosperous farms are reminders of the determination of those settlers who a century ago came to the river to stay.

DAVID MUENCH

***Rising ghostlike above the** cool, blue water of the lake, the giant stone monolith of Temple Bar acts as a landmark for the upper reaches of Lake Mead.*

SUGGESTED READING

ASHBAUGH, DON. *Nevada's Turbulent Yesterdays.* Los Angeles: Westernlore Press, 1963.

CARLESON, HELEN S. *Nevada Place Names.* Reno: University of Nevada, 1974.

DARRAH, WILLIAM CULP. *Powell of the Colorado.* Princeton, Mass.: Princeton University Press, 1969.

MURPHY, DAN. *John Wesley Powell, Voyage of Discovery: The Story Behind the Scenery.* Las Vegas, Nevada: KC Publications, Inc., 1991.

PAHER, STANLEY W. *Nevada Ghost Towns and Mining Camps.* Berkeley: Howell North Books, 1970.

POWELL, JOHN WESLEY. *The Exploration of the Colorado River and Its Canyons.* Reprint of Powell's 1895 *Canyons of the Colorado.* New York: Dover Publications, Inc., 1961.

***Overleaf**: Lights illuminate the roadway and intake towers in the desert dusk from the Arizona approach. Photo by Gail Bandini.*

SNACKS

GIFTS

"It was found that the legal name was— and always had been—Hoover Dam. This continues to be the official name of the dam, although one name is probably as prevalent as the other in popular usage."

Taming the Colorado River

The river, so far, has been portrayed as beneficent and life-giving. And it was. But the Colorado cannot be categorized so easily; there are many facets to its personality. One of these is its tremendously destructive potential.

The Colorado, in its 1,400-mile course from the snow-laden Rockies in Colorado and Wyoming, drains nearly one-twelfth of the total land area in the contiguous United States. Before the 1930s, the spring runoff from this vast area would often converge until the river became swollen and threatening. In Nile-like fashion, it often rose above its banks, depositing millions of tons of silt and flooding hundreds of square miles along its boarders.

The Indians took advantage of the annual floods. As the water subsided, they planted crops in the moist silt deposits. Others also were quick to

GAIL BANDINI

The impossible *became reality with the completion of Hoover Dam in 1935. This dam became the symbol of man's ability to harness nature, even an element as ancient, powerful, and wild as the Colorado River.*

BUREAU OF RECLAMATION

The rugged walls of Black *Canyon, rising nearly 800 feet above the Colorado, presented an engineering challenge the like of which the world had never known. This is the site of the dam in 1931, before construction began. The volcanic rock faces would provide a solid anchor for the great concrete mass that would bridge them at a height of 726.4 feet above the river bed.*

BUREAU OF RECLAMATION

Before actual construction of the dam could begin, *loose rock had to be stripped from the canyon walls. This was the job of the "high scalers," men whose lives depended upon skill, nerve, and heavy-duty suspension ropes!*

That question became the challenge for the government Bureau of Reclamation, established in 1902. Reclamation engineers studied 70 sites that presented possibilities as dam sites. By 1919 the search had narrowed to two sites—Boulder Canyon and Black Canyon—and by 1924 intensive studies had amassed enough data to determine that Black Canyon was the better choice. It would be the site of the most ambitious engineering project the world had yet seen—the harnessing of the mighty Colorado.

The problems that faced the engineers were monumental, but no less so were those facing the lawmakers. The Colorado River drainage basin includes six states. The working out of equitable and mutually acceptable allotments of water among these states and Mexico was no easy task, and it took a number of years to accomplish. The

Herbert Hoover poses for this 1927 photo with Horace Albright, *director of the National Park Service, and Albright's daughter Marian. The Park Service would eventually become a partner of the Bureau of Reclamation in the management of Lake Mead.*

BUREAU OF RECLAMATION

see the potential of the fertile floodplains. By 1901 river water was being diverted via canals into the broad, saucer-shaped Imperial Valley, which lies below the river to the west in California. The low desert valley became a year-round garden.

The farmers, however, had not reckoned with the destructive force of the river. In 1905 it broke through and for nearly two years poured unchecked into the valley, destroying farms, homes, communities, and everything in its path. Levees were built, breached, rebuilt, and breached again. The 300-square-mile Salton Sea is the giant puddle left by that rampage. In other years, however, the quixotic river ran dry bringing drought and miseries of a totally different character.

The cycle of flood and drought became untenable. Something had to be done. But the river was formidable, and it would have its say. Would it allow itself to be tamed?

BUREAU OF RECLAMATION

Amidst intense heat, dust, *noise, and loose rock, work progresses on the floor of Black Canyon. Where the river flowed only a short time before, men and machines undertake a monumental challenge: In less than five years they will construct the largest dam ever built. At the time this photo was taken, the coffer (diversion) dams were in place and the river was flowing through giant tunnels in the canyon walls. Excavations removed centuries of river-deposited alluvium and debris from the surrounding cliffs and provided a firm bedrock base for the dam.*

"Colorado River Compact," initialed by the participating states in 1922, paved the way for the Boulder Canyon Project Act that passed in 1928 (in President Coolidge's administration), which authorized the building of the huge dam in Black Canyon and approved the compact.

The 1928 act provided $175 million for construction, stipulating that the complete cost, except for $25 million allocated as flood-control funds, would be paid back with three percent interest within 50 years. Revenues would be obtained from the sale of electricity generated through the dam's operation over the years. As of October 1, 1976, the project had grossed about $378 million, representing a net return to the goverment of over $202 million, which was applied against repayments of capital costs and interest. The payback has been completed; 1986 saw the final payment and an unusual accomplishment: a government project that paid for itself.

As plans for the dam that would soon rise within its walls progressed, the Black Canyon site became more familiarly known as "Boulder Dam," derived from the Boulder Canyon Project Act's name. This name would stick, in spite of the fact that in 1930, as the result of a suggestion by Secretary of Interior Wilbur, the gigantic structure was named—by Act of Congress—"Hoover Dam." Herbert Hoover had been Secretary of Commerce during the planning years of the dam, and it is largely to his credit that the extremely difficult and intricate negotiations that preceded the signing of the Colorado River Compact and involved six states and Mexico were successful. Had they not been, the building of the dam would have been impossible. It is for this reason—not because Hoover was President of the United States at the time of the 1930 act—that the dam was named for him.

There is a curious sidelight to the matter of the name of the dam, one that perhaps explains the

The dam takes shape, one section at a time, as bucket after *bucket of concrete, each containing eight cubic yards, is swung out over the canyon from an overhead cableway and then poured into the waiting forms. Hoover Dam is a concrete monolith consisting of a series of keyed columns. After each section of a column was poured and the forms removed, grout (liquid concrete) was pumped into the inter-column spaces, forming a one-unit structure of tremendous strength.*

BUREAU OF RECLAMATION

uncertainty surrounding it over the years: The story goes that, when President Roosevelt dedicated the dam in 1935, Secretary of State Harold L. Ickes, irked that the then unpopular (especially among the Democrats) Hoover was being honored, struck the ex-president's name from the dedication speech and substituted "Boulder." This was the name that persisted until President Truman in 1947 asked the 80th Congress to find out just what the name of the dam really was. It was found that the legal name was—and always had been—Hoover Dam. This continues to be the official name of the dam, although one name is probably as prevalent as the other in popular usage.

The engineering problems presented by the gigantic dam had no precedent. The volcanic walls of Black Canyon rose in a sharply vertical angle 800 feet above the surging river confined at the base of the canyon. The site was in an unsettled desert wilderness. The nearest railroad was 40 miles away at the small village of Las Vegas, Nevada, and the nearest source of sufficient electrical power was 222 miles away at San Bernardino, California. Virtually everything necessary to build the structure—including the labor force—would have to be imported.

In 1931 the contract for construction was let to Six Companies, Inc., a consortium of six major western firms. Its successful bid was $48,890,995.50. A townsite was selected and construction of its buildings began. The new town, Boulder City, would house the workers and their families. At the peak of construction, 5,218 people would be working on the dam at one time. A railroad would be constructed by which to haul the astronomical amounts of steel and concrete, and the massive machinery needed to complete the dam.

The procedure for constructing the dam was relatively simple—on paper. Four huge tunnels would be driven into the canyon walls, two on either side, with an opening to the river above and below the site of the dam. Between the tunnel openings, two coffer dams would be built to divert the flow of the river around the construction area. The upstream coffer dam would divert the river into the tunnels. The downstream dam would keep it from flowing back into the construction area. When the river was thus diverted, the construction area would be pumped dry and the tons of water-deposited silt would be excavated down to bedrock. The finished dam—a gigantic, concrete, gravity-arch structure—would soar in a graceful curve between the rugged canyon walls.

In practice the plan posed herculean problems. Nothing of the magnitude of Hoover Dam had ever been attempted in modern history. Problem after problem presented itself, and each one was tackled and solved as it arose, with the help of techniques and equipment that in many cases were highly innovative. Hoover Dam thus became the model for dozens of great dams to follow throughout the world. They were larger dams in many cases, but Hoover Dam was the forerunner of them all. (Once the world's largest hydroelectric installation, it now ranks fifty-fourth.)

The project actually got underway in April of 1931. Men swung out over the canyon on ropes 500 feet above the river bed to place charges of dynamite in the canyon walls that would clear the canyon of loose rock and prepare the site for construction. Clouds of dust rose as the canyon reverberated from the blasts. Load after load of rock and debris was hauled out.

A huge penstock section—a segment of steel pipe 30 feet *in diameter—is lowered into the canyon via a 250-ton-capacity cableway. (The man sitting inside provides scale.) The control house, from which the cableway operator views the procedure above the partially completed dam, is visible at upper left.*

BUREAU OF RECLAMATION

Existing construction equipment was not designed to handle anything of the magnitude of the dam, and in many cases special devices were fabricated on the spot. One of these was the "jumbo," used to drill the diversion tunnels, each 56 feet in diameter, on either side of the canyon. The jumbos were special multiple rock-drilling rigs mounted on 10-ton trucks with a battery of 30 rock drills mounted onto different platform levels.

The trucks were backed up to the end of a tunnel, where water and air lines were connected to them. Working simultaneously, the operators guided all of the 30 bits into the hard rock, some going as deep as 20 feet. The drilled holes were then charged with dynamite, and the resulting blast loosened about 2,400 tons of rock at a time. The gargantuan canyon walls shook in a jittery, stop-and-go dance. Just under a year was required to drive all four tunnels through the canyon walls, 17 feet with each blast. On November 14, 1932, the Colorado River was successfully diverted through the tunnels.

On June 6, 1933, the first bucket of concrete was poured, and the structure rapidly began to take form. Two concrete mixing plants at the site worked day and night. Entire railroad cars loaded with materials were lowered into the canyon by a gigantic tramway. Armies of men, dwarfed by the canyon and the dam, moved in and out in an endless procession, one shift following another. Huge steel conduits for the penstocks, fabricated on the spot yet big enough (30 feet in diameter) to run a railway through, were lowered into place.

Gradually the great columns of concrete took form. Bucket after bucket of concrete was dumped into the waiting forms—8 cubic yards in each, just for the dam alone. The statistics are mind-boggling, including such items as 45 million pounds of reinforcing steel and 5 million barrels of cement.

Even the transporting of the workers was done on a grand scale. Special decks with rows of benches were fitted on the huge ten-ton trucks. From the newly built barracks, houses, and tents at Boulder City, workers were driven 150 at a time down to the dam site in Black Canyon.

Working conditions, for those of us accustomed to the relative quiet and comfort of job situations today, are difficult to imagine. In the summer, the dark rock walls of Black Canyon get so hot they will literally burn the hand that thoughtlessly touches them. The old story of rock being so hot you could fry an egg on it is no exaggeration here. Air temperatures often soar to over 120° Fahrenheit.

The men worked daily, in spite of the heat. Even workers on the night shifts were not completely spared; temperatures rarely dipped below 85° F, even after the sun went down.

These extreme temperatures, together with noise, dust, smoke, and the constant danger from heavy equipment, rocks, and falling debris made accidents an everyday occurrence. A 60-bed hospital was built in Boulder City to handle cases of heat exhaustion and accidents, and two ambulances were always on standby. At the dam site, first-aid

D*avis Dam, named for former Bureau of Reclamation Director* *Arthur Powell Davis, was completed in 1953. The 200-foot-high, earth-and-rockfill dam with concrete spillway is a multipurpose structure that generates hydroelectric power at the rate of over 1 billion kilowatt-hours per year for citizens of southern Nevada, Arizona, and California. Davis Dam forms Lake Mohave, backing the Colorado River upstream 67 miles, almost to the base of Hoover Dam.*

GAIL BANDINI

stations were set up that handled as many as 1,500 minor injuries per month. (These hardy workers considered as "minor" almost any injury that did not result in death.)

It is a tribute to the planning and organization of the project that there were as few serious injuries and deaths for the entire project as there were, given these arduous and dangerous conditions. Ninety-four construction workers and 2 government employees were killed between 1931 and 1937: 24 fell to their deaths, 3 drowned, 10 were killed by explosions, 5 were electrocuted, 26 were struck by falling debris, 26 were struck by machinery or heavy equipment, 1 died in an elevator accident, and 1 died in a cableway accident.

Heat was also an enemy in the pouring of the concrete. The heat generated by concrete as it cures in volumes such as those that were to be poured in the huge dam would take many, many years to dissipate. To prevent this, the dam was poured in a series of 230 interlocked columns of blocks, each 25 to 60 feet in width. Water, cooled to 38° F, was then circulated through tubing to help cool the concrete. The water was cooled by a 3,000-gallon-per-minute cooling tower and refrigeration plant capable of producing 1,000 tons of ice per day. Thus, the entire structure was cooled in 20 months rather than the 150 years it was estimated it would have taken without the aid of the refrigerated-water process.

The most incredible statistic of all was the length of time it took to complete the project—a period of *less than five years*. Five years to accomplish an engineering feat that would astound the entire world!

On February 1, 1935, the tunnels diverting the water were closed at the intakes and Lake Mead began to form. Cooling was completed in March and the last concrete was poured May 29. On September 30, 1935, President Franklin D. Roosevelt officially dedicated the dam in ceremonies appropriate to the occasion. On October 22, 1936, the first generator began producing power, with a second going into operation on November 14 and a third on December 28. The river, which had been master and creator of this vast region for eons, had finally bowed to man.

Hoover Dam Today

The statistics concerning the completed dam and the Lake Mead reservoir are as mind-boggling as those concerning its construction. The lake has a capacity of over 32 million acre-feet of water when full—enough water to cover the entire state of New York to a depth of 1 foot. It will hold the entire average flow of the river for two years. When full it reaches, in places, a depth of about 500 feet. The lake is 110 miles long.

The dam itself is a gigantic 726.4 feet from the bedrock of Black Canyon to the crest of the dam. It is 660 feet thick at the base. The dam, powerhouse, and related structures contain over 4,360,000 cubic yards of concrete. The intake towers that divert water into the turbines stand as high as a 34-story building.

The last generator (17th) began operation on December 1, 1961, bringing the power plant capacity to over 2,000,000 kilowatts. In its first 40 years of commercial power generation, Hoover Dam produced approximately 150 billion kilowatts—enough energy to supply a million residences for 20 years. Generation of this amount of energy in an oil-burning plant over the same time would have consumed approximately 258 million barrels. The Bureau of Reclamation is at present examining possibilities for increasing the generating capacity of the power plant in order to improve its ability to meet peaking demands on the systems it serves.

Passing in front of *Hoover Dam, the Thunderbirds—flying F-16 Fighting Falcons—execute one of the many tight-formation maneuvers for which this superbly coordinated team is famous. The U.S. Air Force Air demonstration Squadron operates out of nearby Nellis Air Force Base in Las Vegas, Nevada. For over 40 years the Thunderbirds have routinely performed more than 65 shows annually for people all around the world, winning friends and recruits for the Air Force with their remarkable skills.*

PHOTO BY SSGT JACK BRADEN, COURTESY U.S. AIR FORCE AIR DEMONSTRATION SQUADRON, THE THUNDERBIRDS

DAVIS DAM

Other dams followed Hoover, both upstream and downstream. Davis Dam, 67 miles downstream, was begun in 1942, but work was halted in November because of materials shortages brought on by the war. Work was resumed in 1946 and the dam was completed in 1953, fulfilling an obligation that the United States incurred in the Mexican Water Treaty of 1944. Its reservoir, Lake Mohave, provides storage to meet the requirements of delivery of a certain portion of the river's water to Mexico. This 200-foot, earth-and-rockfill dam also generates hydroelectric power for southern Nevada, Arizona, and California.

Today, except where it runs through the Grand Canyon, the Colorado in its lower basin is controlled and regulated by a network of dams along its entire course, from Lees Ferry southward. Flood control, generation of electricity, and reliable water supplies for agricultural and domestic use for several million people are the impressive and obviously measurable results. Less easily quantified but perhaps equally important are the recreational opportunities the various reservoirs supply for millions of people.

Administered by the Bureau of Reclamation, Hoover Dam and Davis Dam—together with their reservoirs at Lakes Mead and Mohave—continue to fulfill and expand their intended purposes. With the filling of these great water bodies, the lakes, a new era began in man's ever-changing relationship with the river.

SUGGESTED READING

Construction of Hoover Dam. Facsimile of a revision of the 1936 edition, Bureau of Reclamation information booklet on the engineering of the dam. Las Vegas, Nevada: KC Publications, Inc., 1976.

The Story of Hoover Dam. U.S. Dept. of Interior. Washington, D.C.: U.S. Gov't Printing Office, 1971.

From Water to Electricity – A Masterpiece of Engineering

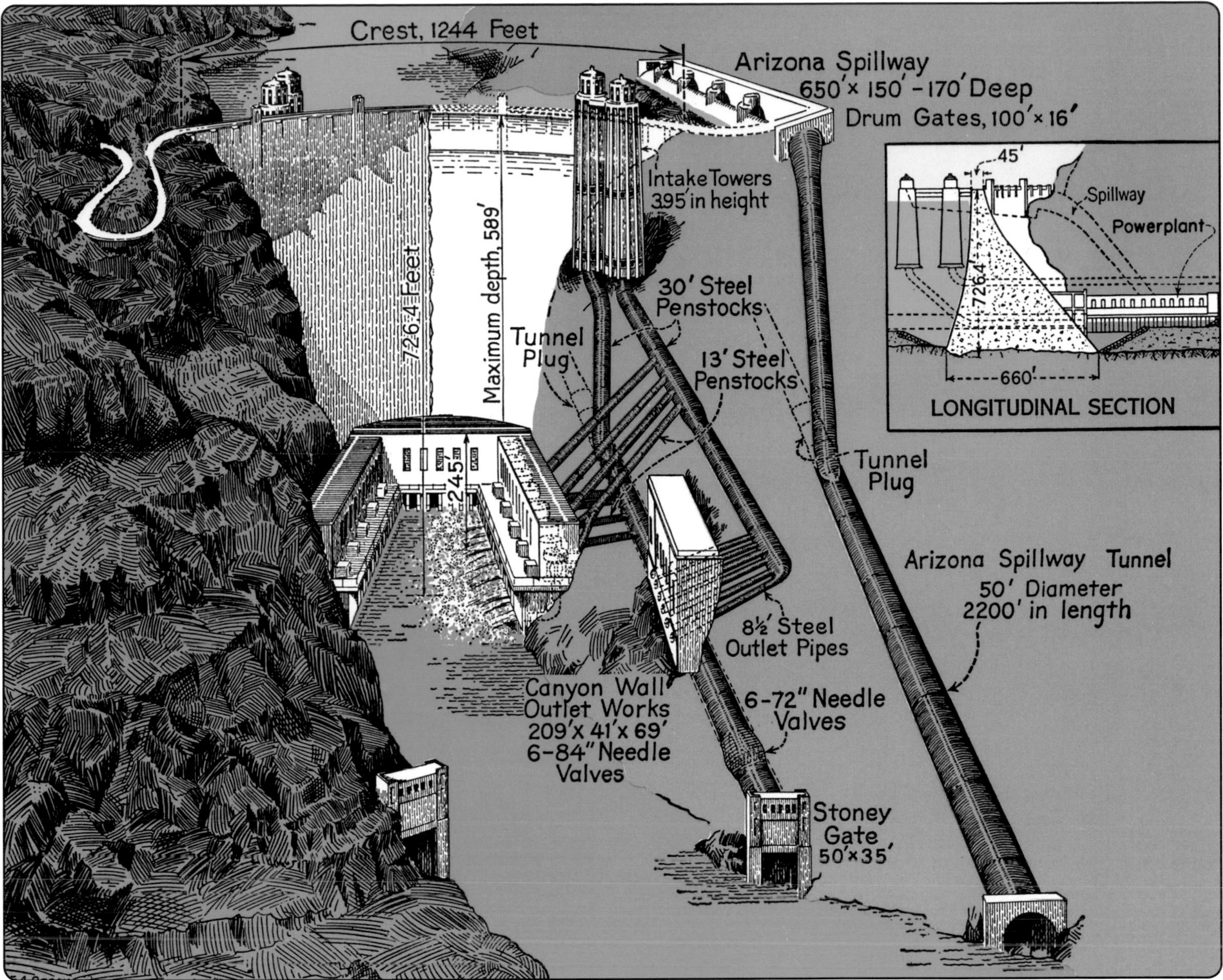

COURTESY BUREAU OF RECLAMATION

Visitors who tour the dam can see the *banks of huge electric generators that are located in the powerhouse wings on either side of the canyon below the dam. The water, under tremendous pressure, flows in 13-foot-wide penstocks through the turbines—devices that might be described as very sophisticated "water wheels." The spinning turbines are connected to the steel shafts that drive the generators. The 17 generators at the dam have a combined capacity of 2,080,000 kilowatts of electric energy.*

K.C. DEN DOOVEN

The engineers of Hoover Dam anticipated almost every eventuality and made provisions that ranged from prestressing cold-water intake pipes to installing a system that could handle the highest lake level imaginable. Unlike many dams, Hoover Dam channels water around the structure, instead of over or through it. Drawn from the depths of Lake Mead, water enters the four intake towers, flows through the canyon walls (on both the Nevada and Arizona sides) in steel penstocks into hydroelectric generator turbines, and then rejoins the river. To prevent abnormally large volumes of water from ever overflowing and damaging the dam, battleship-sized spillway basins were constructed at maximum-water level on each side. In this unlikely event, water would enter the spillways, drain through 50-foot-wide tunnels, and enter the canyon well below the dam. Normal flows, obviously, are only a fraction of the total capacity of this outlet system, an engineering feat that is truly impressive, even today.

"Lake Mead and Lake Mohave offer the visitor identical opportunities for water recreation. Yet each lake has its own qualities and its own special appeal."

Lakes Mead and Mojave Today

As Lake Mead was filling, it became apparent that here would be a unique resource. A giant lake in the desert would offer almost unlimited water-based recreation on a year-round basis. The U.S. Bureau of Reclamation, realizing the magnitude of the job of administering such a vast resource, turned to its sister agency in the Department of the Interior for its expertise and help. By a memorandum of agreement, the National Park Service assumed administration of "Boulder Dam Recreation Area" on October 13, 1936.

GAIL BANDINI

The new Hoover Dam visitor center—a three-level, *110-foot diameter circular structure with a rooftop overlook—opened to the public in 1995. The middle level of the building houses a rotating theater divided into three 145-seat sections. Adjacent to the center is a tour elevator shaft, enclosed by a faceted glass tower.*

These striking sculptures—30 feet high and *containing more than 4 tons of statuary bronze—are probably the largest monumental bronzes ever cast in the United States. They rest upon a base of California black diorite set into a terrazzo floor inlaid with a "star chart," from which a 142-foot flagstaff rises. Sculptor Oskar J. W. Hansen explains that these* Winged Figures of the Republic *are "mighty of body and clean of soul, armed only in the winged imagination of their own thoughts," and therefore symbolize the "mental fire, daring, and imagination" of the builders of Hoover Dam. The figures also have the clear piercing "look of eagles," which symbolizes that "eternal vigilance . . . is the price of liberty."*

GAIL BANDINI

GRAY LINE
GRAY LINE
TOURS

***Boating is certainly the** most popular way to see the recreation area! Many people spend more than one day at the lakes, enjoying nights under clear, starry skies in the still desert. Lakeshore camping is permitted at most locations. Other boaters bring their "houses" with them.*

GAIL BANDINI

***This handsome catch, striped bass, is a popular** game fish. The lakes reward many anglers, whether they fish fromshore or from boats.Government stocking programs have also benefited shorebirds, which lake their share of the fish.*

K.C. DEN DOOVEN

With this agreement, a new type of unit was added to the growing national park system—the national recreation area—in recognition of the fact that areas that offered outstanding outdoor recreational opportunities were a national resource worthy of preservation. As the new recreation area took its place among the great natural and historic shrines of America, a new concept emerged in the history of conservation.

The national parks, monuments, and historic areas of the National Park Service had since 1916 provided opportunities to learn and to appreciate the natural and cultural heritage of America. The recreation area, however, gave visitors an outdoor experience of a different sort. Others would follow, but few of these would offer opportunities for recreation as varied and activity-oriented as Lake Mead.

The complications of administering such an area created many new challenges that, like the engineering challenges of the dam itself, had to be promptly met. As a result, much of the basic foundation of policy and philosophy used in the operation of national recreation areas all over the United States was developed at Lake Mead.

On October 8, 1964, President Lyndon Johnson signed the act that formally established the "Lake Mead National Recreation Area." This act redesignated the old Boulder Dam Recreation Area, whose boundaries had been substantially enlarged in 1947 to include the yet-to-be-filled Lake Mohave, in recognition of its equally significant recreational opportunities. The area as established in 1964

GAIL BANDINI

Many marinas have rental boats *available. Here at Las Vegas Bay on Lake Mead, visitors can obtain fishing and pontoon boats, as well as powerboats for water-skiing and exploring. Some marinas even offer houseboats that sleep up to 12 people.*

included some 3,000 square miles of lake and desert. The eastern boundary was the western boundary of Grand Canyon National Monument. In actuality, then, the recreation area encompassed over 90 miles of the westernmost Grand Canyon. It also included the highland area north of the Grand Canyon known as the "Shivwits Plateau."

In 1974 the boundaries of the recreation area were again modified. Grand Canyon National Park was expanded to include all of Grand Canyon National Monument and the Lake Mead portion of the Grand Canyon. Thus the entire Grand Canyon came to be under one administrative unit. The Shivwits Plateau was included in the Grand Canyon-Parashant National Monument in January 2000.

The muddy waters of the river are now calm. For most visitors the essence of Lake Mead National Recreation Area is in fact the clear, clean lakes. Water is king and its reign is largely a happy one.

Lake Mead and Lake Mohave offer the visitor identical opportunities for water recreation. Yet each lake has its own qualities and its own special appeal.

Lake Mead

Lake Mead is by far the larger of the two lakes. Boulder Canyon divides the lower—or Boulder—basin from the Virgin Basin, the Virgin Arm, and the smaller upper basins. Hoover Dam backs up water into the lower valley of the Virgin River for 25 miles, forming a third large basin.

The open expanses of water in these three large basins characterize Lake Mead. There's room here—marvelous room—for sailing, water skiing, cruising, and just communing with nature on a vast

Thousands of boaters use the lakes each year, but *they are not crowded. The more than 700 miles of shoreline offer countless opportunities for exploration. One can return again and again to a favorite cove or hideaway in which to enjoy the special solitude that can only be found here, where water and desert contrast and complement each other in an immensely satisfying way.*

K.C. DEN DOOVEN

***Powerful jet-propelled** racers roar across the lake, adding turbulence to the waves. Lake Mead is often the scene of water-racing events that range from sedate sailboat races to long-distance water-skiing.*

GAIL BANDINI

scale. In addition, the countless side canyons that have been flooded by the lakes provide seemingly unlimited numbers of coves for exploring, fishing, swimming, camping, and getting to know nature—and thereby oneself—a little better. In all, nearly 550 miles of shoreline have been created by these side canyons.

The spectacular scenery of Iceberg and Virgin canyons in the upper reaches of Lake Mead provide a special backdrop for boaters in that part of the lake. The launching facilities at South Cove are a starting point for cruises into the lower Grand Canyon. At the higher lake levels, it is possible to take powerboats into the Grand Canyon, provided proper precautions against sandbars and driftwood are taken. A trip to the base of Grand Wash Cliffs (where the Colorado emerges from the Grand Canyon) and then up into the Grand Canyon itself is a never-to-be-forgotten experience.

Although much of Lake Mead must be experienced by boat, the various campgrounds, marinas, lodges, and other facilities clustered around the lake make it possible for non-boaters to enjoy it as well. Literally millions of people use the lake each year. And many of these visitors return again and again—to find that special cove or campground, or just to sit on the banks and enjoy solitude of a quality that only nature can supply.

GAIL BANDINI

***The** Desert Princess, a Mississippi-style paddle-wheeler, cruises Lake Mead several times daily. Breakfast, dinner and dance, and special events trips are available, with schedules varying according to the season. Hoover Dam tour tickets may be included with any daytime cruise.*

GAIL BANDINI

A series of concessionaire-operated *marinas are strategically located on the shores of Lakes Mead and Mohave. Most of them, such as the Callville Bay Marina, Lake Mead, have stores, restaurants, and campgrounds, as well as facilities for launching, berthing, and servicing boats.* ***As the flow of the Colorado River changes****, the levels of the lakes change too–often dramatically. At times all the marinas look very different.*

Lake Mead—named for Elwood Mead, the Bureau of Reclamation commissioner who was in charge of *project activities during most of the construction—is a place where one can revel in the excitement, beauty, and exhilaration of sparkling water and warm sun. In contrast to the isolated coves that can be reached only by boat, Boulder Beach is easily accessible; it draws thousands of people every year.*

GAIL BANDINI

Long and narrow, Lake Mohave *retains much of the feeling of the river. Between the confining walls of Black Canyon, the lake is not much wider than the river was when it flowed freely through.*

GAIL BANDINI

"...desert **bighorn** **feed** on lakeshore vegetation while **shorebirds** **fish** the **lake** nearby."

Lake Mohave

In contrast, Lake Mohave is smaller and more intimate. It retains much of the character of the river. Narrow canyons cut by the Colorado confine it for much of its 67-mile length from Davis Dam upstream to a point near the base of Hoover Dam. The canyons fall away near the middle of Lake Mohave. In Cottonwood Valley, the Cottonwood Basin expands to approximately 4 miles wide, the largest open part of the lake.

Parts of Lake Mohave, however, are no wider now than when the river was freely flowing. In fact, the river, at the upper end of Black Canyon, still flows over rapids. This occurs only during periods of low-lake levels. (One or two of the old ringbolts used by the steamboat captains remain in the canyon walls here, offering mute testimony to the struggle that occurred here between the paddle-wheelers and the river.)

Lake Mohave also has its share of beach camps, superbly situated coves to explore, and commercial areas that offer a full complement of camping, lodging and boating facilities.

Both lakes are internationally known sport fisheries. Creation of the lakes has allowed the introduction of several prized species of game fish. Large-mouth bass has been the principal game fish since the filling of the lakes, and anglers also have had the opportunity of taking record-sized rainbow trout in the cold waters of Black Canyon on Lake Mohave. The U.S. Fish and Wildlife Hatchery at Willow Beach on Lake Mohave in Arizona operates a year-round trout-stocking program for Lake Mohave. The Lake Mead Fish Hatchery of the Nevada Fish and Game Department has stocked channel catfish, black crappie, perch, and striped bass, which are now self-supporting.

The lakes attract an unusual variety of animals, at least for the desert. Cormorants, ducks, geese, pelicans, herons, gulls, and other water-oriented birds share the lakeshore scene with desert animals in a unique blending of wildlife that is a normal sight here. In Black Canyon, for instance, desert bighorn feed on lakeshore vegetation while shorebirds fish the lake nearby.

It is not surprising that the majority of visitors to Lake Mead National Recreation Area have been here before. Perhaps it is the wide variety of sights, activities, and experiences that compel people from all over the country to return to the lakes again and again.

GAIL BANDINI

Black Canyon Raft Tours offers three- *hour float trips down the Colorado River. Passengers board 40-person rafts below Hoover Dam and travel the current to Willow Beach. Individuals can rent smaller boats at Willow Beach and travel either upriver or downriver.*

GAIL BANDINI

Katherine Landing is the *southernmost marina in the recreation area. In addition tofull marina services, Katherine Landing offers numerous options for spending the night: a campground,a trailer village (with hookups), or even a hotel—one of five on the lakes.*

Visitors come because of the water, but they are often surprised to discover that the desert has beauty and value just as rewarding. The water provides opportunities for recreation and relaxation, and the desert offers a learning experience in the unique adaptions that plants and animals have had to make to their often-harsh surroundings. Summertime visitors may find little reason to stray far from the water or from their air-conditioned autos, but when the intense heat lessens by late September, the desert becomes a pleasant and inviting place where one can hike, camp, take photos, or simply relax in the now hospitable sun.

Like the river, the desert has many sides to its personality and cannot be taken for granted. Flash flooding, as the term implies, can happen here in an instant. An example was the 1974 flooding of the small village known as Nelson's Landing at the mouth of Eldorado Canyon, where people had lived for nearly seven decades without incident. The desert gave no warning. And it gave no mercy. Many lives were lost and homes swept away in the sudden deluge that followed a summer cloudburst.

But for the most part, life goes on in the desert with few interruptions, especially of this severity. It is almost startling to learn that, far from a lack of life here, the desert offers a relative abundance of living things. Through countless generations of natural selection, each successful species has developed some means of solving the problem of obtaining and retaining moisture in sufficient quantities to maintain life. Thus, a variety of animals and plants thrive under seemingly impossible conditions.

Barring extreme interference by man, nature maintains a delicate but constant balance, a web of life that is made up of the various plants and animals and resources that sustain them. Desert conditions are not static; they change constantly. Landforms are even now undergoing continuous, if imperceptible, reshaping; climate, too, is continuously being modified. But a balance must always be maintained. This simple rule applies almost anywhere in the natural world, but in the desert it is particularly important. In this parched land all life seems preoccupied with the processes of obtaining and retaining water. The intense heat and direct sunlight hasten loss of water from plants and animals through transpiration and perspiration.

The unusual forms that many of the plants in the Lake Mead country take are a result of the adaptations. Most have tiny leaves, so as to expose the least area to water loss through transpiration. The small leaves of the creosote bush, the dominant plant in the Lake Mead area, have a resinous coating that also cuts down water loss.

Cacti are plants that have all but lost their leaves. Green, waxy pods and stems perform the necessary functions of photosynthesis that in other plants are carried out by leaves. At the same time the waxy coating helps conserve water in the built-in water-storage reservoirs of the cacti. Most desert plants have extensive shallow-root systems that allow them to take in the maximum amount of moisture from the infrequent rains.

The annuals, which add so much color to the spring desert, germinate, grow, mature, flower, and produce seed during the few weeks that moisture from the late-winter and early-spring rains is relatively abundant. The seeds then lie dormant until water is again available. Moisture-loving plants, such as mesquite and catclaw, may send taproots

GAIL BANDINI

The desert often responds to winter rains with generous bursts *of springtime life and color. In just a few short weeks, dozens of plant species germinate,grow, and blossom. Here, beavertail cactus and desert trumpet highlight the view toward the lake.*

GAIL BANDINI

***V*isitors are thrilled by the** *huge white blossoms of the Mohave yucca, but this plant is also quite beneficial. Indians utilized every bit of the plant. They made rope from its long fibrous leaves, flower petals were eaten fresh, and seeds were ground into meal. Birds and small mammals also use the seeds as an important food source.*

GAIL BANDINI

***T*he silvery spines of the teddy bear** *cholla look cuddly, but beware—passersby often learn how its loose joints earned it the nickname "jumping cholla."*

down as much as 60 feet to reach the water table in the seemingly dry desert washes.

Animals, too, exhibit remarkable adaptations, often through modifications in behavior from season to season. During the summers, most become nocturnal, feeding or hunting during the cooler night hours and sleeping in underground burrows or shaded areas during the day. This is particularly true of reptiles, small mammals, and many birds. The desert bighorn, however, seems to successfully defy the searing summer heat, browsing at midday and going for several days at a time without taking a drink.

With the coming of the lakes, obvious changes in desert conditions have occurred along their shorelines: Water-adapted birds have found the lakes. Several of the fishes native to the warm, muddy water of the Colorado River have apparently disappeared in the clear, still, man-made lakes; new, introduced species have replaced them. The relatively abundant vegetation that once lined the banks of the river and was flooded by the rising waters of the lakes has not reestablished itself in the relatively short period of the existence of the lakes.

This is due in large measure to the fluctuation of water levels. But it must be borne in mind that a dynamic period of adjustment between lake and desert is still taking place here. What is surprising, however, is how little the lakes have affected the basic desert patterns. Only a few yards back from the water, the vast desert impassively resists change. This filling of the lakes has affected only the small part of the desert that the lakes have touched with their shorelines.

SUGGESTED READING

EVANS, DOUGLAS B. *Auto Tour Guide to the Lake Mead National Recreation Area.* Globe, Arizona: Southwest Parks & Monuments Association, 1971.

HOUK, ROSE. *Lake Mead National Recreation Area.* Tucson, Arizona: Southwest Parks & Monuments Association, 1997.

ROHDE, KATHERINE M. *in pictures Lake Mead: The Continuing Story.* Las Vegas, Nevada: KC Publications, Inc., 1999.

All About Lake Mead National Recreational Area

Western National Parks Association

Western National Parks Association (WNPA), was established in 1938 as a nonprofit organization, that works with the National Park Service to increase visitor understanding and appreciation of national parks throughout the West.

It provides support and funding for research, education, and interpretation in the recreation area. Several free handouts and the visitor newspaper are funded by WNPA. It has assisted in funding new map exhibits in eight information centers and supported the design and production of new signs for the recreation area's cactus gardens.

Beavertail cactus
Photo by Gail Bandini

For information about the park:

Write to:
Lake Mead NRA
601 Nevada Highway
Boulder City, NV 89005

Park information #
702-293-8990

Fax #
702-293-8936

Website:
www.nps.gov/lame

If you would like to volunteer at Lake Mead Recreational Area please contact the Volunteer Coordinator at:
702-293-8714.

Lake Mead & Hoover Dam

Junior Ranger

JUNIOR PARK RANGER
LAKE MEAD NRA
NATIONAL PARK

Become a Lake Mead Recreational Area Junior Ranger! You will discover the exciting world of the plants and animals that live in and around Lake Mead. Discover why it is important that you "Leave no Treads," and take care of the countryside around you. How can you protect everything within and outside of the park? Ask a Park Ranger and he/she will tell you! Learn what different Park Rangers there are and what jobs they do. Become a Junior Archaeologist and discover who lived here long ago and what they left behind.

Pick up a Junior Ranger Guide and Activity booklet from the Alan Bible Visitor Center or Katherine Landing Ranger Station. Complete the activities within, and get started on an adventure of a life time!

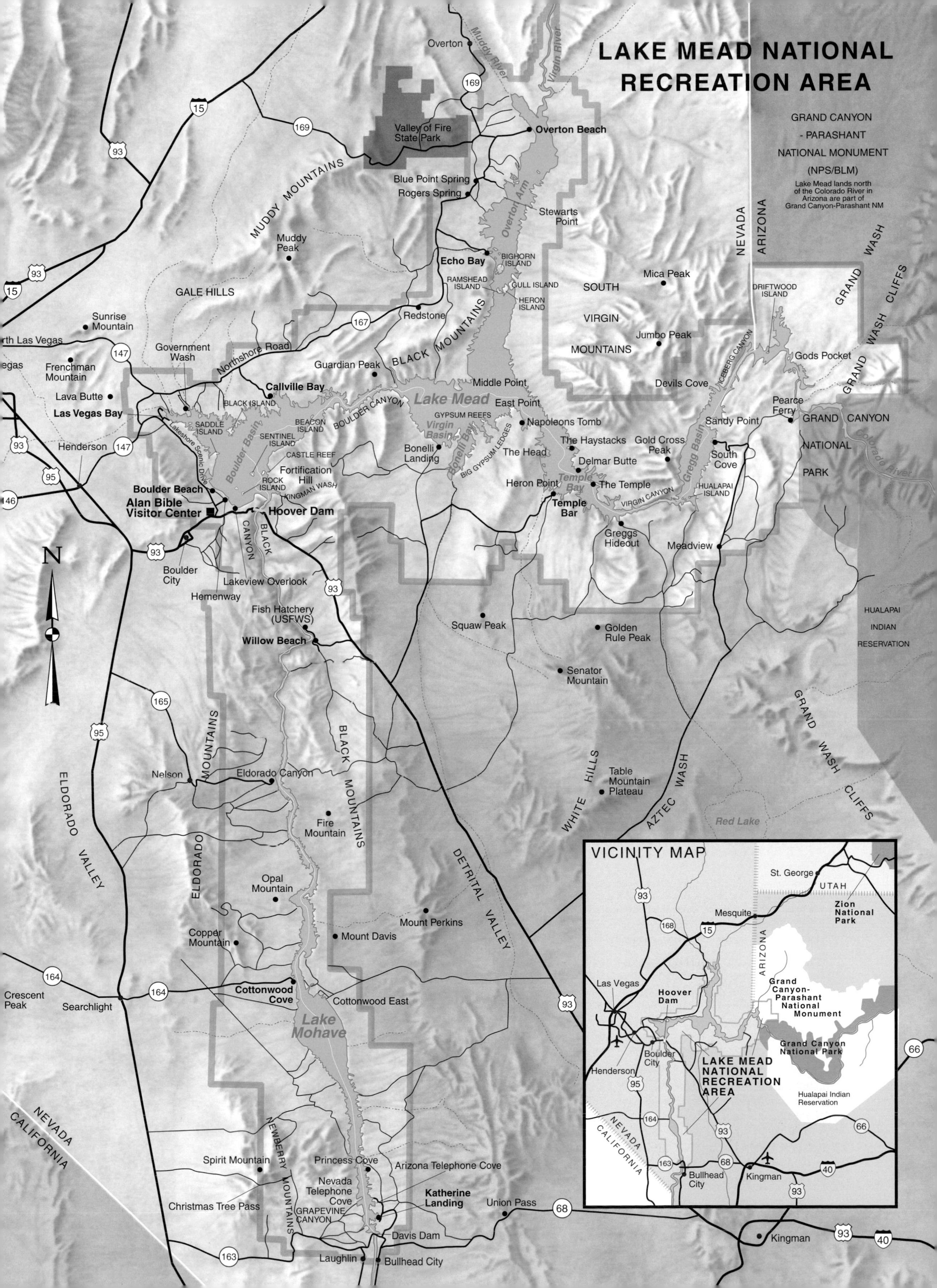

LAKE MEAD NATIONAL RECREATION AREA
GRAND CANYON - PARASHANT NATIONAL MONUMENT (NPS/BLM)
Lake Mead lands north of the Colorado River in Arizona are part of Grand Canyon-Parashant NM
Overton
Muddy River
Virgin River
Valley of Fire State Park
Overton Beach
Blue Point Spring
Rogers Spring
Stewarts Point
Overton Arm
MUDDY MOUNTAINS
Muddy Peak
Echo Bay
BIGHORN ISLAND
RAMSHEAD ISLAND
GULL ISLAND
HERON ISLAND
GALE HILLS
Sunrise Mountain
North Las Vegas
Frenchman Mountain
Government Wash
Northshore Road
Redstone
BLACK MOUNTAINS
Guardian Peak
Lava Butte
Callville Bay
Las Vegas Bay
BLACK ISLAND
SADDLE ISLAND
Lakeshore Scenic Drive
Boulder Basin
SENTINEL ISLAND
BEACON ISLAND
BOULDER CANYON
CASTLE REEF
Fortification Hill
ROCK ISLAND
KINGMAN WASH
Henderson
Boulder Beach
Alan Bible Visitor Center
Hoover Dam
BLACK CANYON
Boulder City
Lakeview Overlook
Hemenway
Lake Mead
Middle Point
East Point
GYPSUM REEFS
Virgin Basin
Bonelli Bay
BIG GYPSUM LEDGES
Bonelli Landing
Napoleons Tomb
The Haystacks
The Head
Delmar Butte
Heron Point
Temple Bay
The Temple
Temple Bar
VIRGIN CANYON
Greggs Hideout
Gold Cross Peak
SOUTH VIRGIN MOUNTAINS
Mica Peak
Jumbo Peak
Devils Cove
NEVADA
ARIZONA
DRIFTWOOD ISLAND
ICEBERG CANYON
Gregg Basin
Gods Pocket
Sandy Point
South Cove
Pearce Ferry
HUALAPAI ISLAND
Meadview
GRAND WASH CLIFFS
GRAND CANYON NATIONAL PARK
Colorado River
HUALAPAI INDIAN RESERVATION
N
Fish Hatchery (USFWS)
Willow Beach
Squaw Peak
Golden Rule Peak
Senator Mountain
ELDORADO MOUNTAINS
Nelson
Eldorado Canyon
BLACK MOUNTAINS
Fire Mountain
ELDORADO VALLEY
Opal Mountain
Copper Mountain
Mount Davis
Mount Perkins
DETRITAL VALLEY
WHITE HILLS
Table Mountain Plateau
AZTEC WASH
GRAND WASH CLIFFS
Red Lake
Crescent Peak
Searchlight
Cottonwood Cove
Cottonwood East
Lake Mohave
NEVADA
CALIFORNIA
Spirit Mountain
NEWBERRY MOUNTAINS
Christmas Tree Pass
Princess Cove
Arizona Telephone Cove
Nevada Telephone Cove
GRAPEVINE CANYON
Katherine Landing
Union Pass
Davis Dam
Laughlin
Bullhead City
Kingman
15 93 95 146 147 163 164 165 167 169 68 40 66
VICINITY MAP
St. George
UTAH
Zion National Park
Mesquite
ARIZONA
Grand Canyon-Parashant National Monument
Las Vegas
Hoover Dam
Grand Canyon National Park
Boulder City
Henderson
LAKE MEAD NATIONAL RECREATION AREA
Hualapai Indian Reservation
NEVADA
CALIFORNIA
Bullhead City
Kingman

Nature's Heritage

Hoover Dam and Davis Dam harnessed the Colorado, but merely controlling its force did not break the spirit of the river. It is this spirit, together with the exhilaration and challenge of the environment, that consciously or unconsciously lures millions of visitors each year. Once here, many find that the fun and excitement of the lakes—skiing, fishing, and sailing—may be eclipsed by the powerful appeal of the vast and mysterious desert. Raw and harsh, yes, but it has a subtle beauty that is almost hypnotic in its pull. Beyond the exciting memory of subduing a fighting fish lie the equally satisfying memories of the desert—a magnificent desert bighorn fleetingly glimpsed on the cliffs above the lake, a brilliant desert sunrise spreading over the waters, and layers and layers of purple mountains that stretch to the late-afternoon horizon.

River, desert, mountains, and the man-made lakes and dams—all speak of power and beauty on a scale so grand that it must be personally experienced to be believed. Thanks to its preservation as a unit of the National Park Service, we are assured that it will speak as compellingly to millions of our sons and daughters in the generations to come as it does to us. The area is an island of serenity that we can count on to remain relatively the same in a world where phenomenal change is increasingly the rule. Let us cherish and preserve it as the important, indeed *vital*, part of our heritage that it is.

ED SCOVILL

Desert bighorn, sailboats, and barren hills symbolize the versatility of the Lake Mead country.

Published by KC Publications, 3245 E. Patrick Ln., Suite A, Las Vegas, NV 89120.

Inside back cover: *Sunrise softens rugged peaks opposite Willow Beach. Photo by Gail Bandini.*

Back cover: *Colorful spring desert flowers frame the blue waters of the man-made lake. Photo by Gail Bandini.*

Created, Designed, and Published in the U.S.A.
Printed by Tien Wah Press (Pte.) Ltd, Singapore
Pre-Press by United Graphic Pte. Ltd